# KEY TEXTS

THOEMMES

Printed and Bound by
Antony Rowe Ltd., Chippenham, Wiltshire

**Classic Studies in the History of Ideas**

# THE UNITY OF SCIENCE

## Rudolf Carnap

Translated with an Introduction by
**M. Black**

© Thoemmes Press 1995

Published in 1995 by
Thoemmes Press
11 Great George Street
Bristol BS1 5RR
England

ISBN 1 85506 391 3

This is a reprint of the 1934 Edition

Publisher's Note

These reprints are taken from original copies of each book. In many cases the condition of those originals is not perfect, the paper, often handmade, having suffered over time and the copy from such things as inconsistent printing pressures resulting in faint text, show-through from one side of a leaf to the other, the filling in of some characters, and the break up of type. The publisher has gone to great lengths to ensure the quality of these reprints but points out that certain characteristics of the original copies will, of necessity, be apparent in reprints thereof.

# CONTENTS

|  | PAGE |
|---|---|
| INTRODUCTION BY M. BLACK | 7 |
| AUTHOR'S INTRODUCTION | 21 |
| PHYSICS AS A UNIVERSAL SCIENCE | |
| 1. THE HETEROGENEITY OF SCIENCE | 31 |
| 2. LANGUAGES | 37 |
| 3. PROTOCOL LANGUAGE | 42 |
| 4. THE PHYSICAL LANGUAGE AS AN INTERSUBJECTIVE LANGUAGE | 52 |
| 5. THE PHYSICAL LANGUAGE AS A UNIVERSAL LANGUAGE | 67 |
| 6. PROTOCOL LANGUAGE AS A PART OF PHYSICAL LANGUAGE | 76 |
| 7. UNIFIED SCIENCE IN PHYSICAL LANGUAGE | 93 |

# INTRODUCTION

BY M. BLACK

1. *Origins of the Viennese Circle*[1]. The so-called Viennese circle of philosophers, to which Professor Carnap belongs, inherits a tradition of empirical and antimetaphysical thought, continuous in Vienna since the middle of the Nineteenth Century and fostered by a long series of eminent university teachers of Philosophy and Science. The growth of this trend of thought coincided with the rise of Liberalism in Austria-Hungary (especially from 1848 onwards) and drew much of its inspiration from the empirical and utilitarian elements of progressive thought in England at the same period. Of the academic teachers prominently in sympathy with this movement, some of the

[1] For the subject-matter of this section I am chiefly indebted to a pamphlet, " Die wissenschaftliche Weltauffassung. Der Wiener Kreis ", in the series *Veröffentlichungen des Vereines Ernst Mach* (Vienna, 1929). The same pamphlet also contains a useful detailed bibliography.

best known are Th. Gomperz (1869–80)[1], the translator of J. S. Mill, Mach (Privatdozent 1861–4, Professor 1895), and Boltzmann (Mach's successor, 1902–6). Parallel to the work of these men were the attempts made to reform the traditional Logic of Aristotle and the Scholastics, of which the beginnings can be seen in Bolzano (especially in the *Wissenschaftslehre*, 1837), and a fuller development in Brentano (Professor of Philosophy in the Theological Faculty, 1874–1880, afterwards Dozent in the Philosophical Faculty) and Höfler (1853–1922). Of the many who took an active part in the philosophical discussions of Brentano's circle in the late Nineteenth Century, we may pick out von Meinong (in Vienna 1870–82, afterwards Professor at Graz).

Subsequently, the most effective influences on this trend of thought were the researches, in Logic and the Foundations of Mathematics, of Russell and others of the ' logistic ' school[2] (especially through *Principia Mathematica*, 1910). Russell's influence has been since reinforced by

[1] The dates given refer in each case to the period spent in Vienna.
[2] Cf. M. Black, *The Nature of Mathematics*, p. 7 and pp. 15 ff. for an account of these theories.

Wittgenstein's *Tractatus Logico-Philosophicus* (published in book-form in 1922) which solved one of the major problems of an empiricist outlook by providing a more satisfactory solution of the nature of Logic and Mathematics. This book has been the chief inspiration of many distinctive features of the contemporary positivist[1] movement in Vienna.

The 'Viennese circle', in its present form, originated in informal discussions dating from the appointment of Moritz Schlick, in 1922, as Professor of Philosophy in Vienna. Out of these arose the " Verein Ernst Mach " (formally created in 1928 with Schlick as chairman), an association " for propagating and furthering a scientific outlook ", and for " creating the intellectual instruments of modern Empiricism ". The first number of *Erkenntnis*, a periodical devoted to furthering the aims of the circle[2], appeared three years ago (1931). The long article by Professor Carnap which has been translated

[1] This and other descriptions occurring in the above are misleading if taken too literally. Cf. page 29 below.

[2] *Erkenntnis* contains reports of the various conferences, lectures, etc., arranged by the group, in addition to papers on Philosophy and the methodology of Science.

for this book appeared in *Erkenntnis* (Vol. ii, 1932, pp. 432-465) under the title " Die physikalische Sprache als Universalprache der Wissenschaft ", and has been revised, by the author, for this edition.

An authoritative statement[1] of the programme of the Viennese circle declares that its outlook is " characterized, not so much by special assertions, as by its fundamental attitude, its point of view and by the direction of its researches. Its goal is unified Science : its endeavours are to relate and harmonize achievements of individual researchers in the various branches of Science. From this choice of subjects arises the emphasis on collective work ; hence also the prominence allotted to communicable knowledge ; these aims inspire the search for a neutral system of symbols, free from the dross of historical languages, the search for a complete system of concepts. We strive for order and clarity, reject all dim vistas and fathomless depths. In Science there are no ' depths ', all is on the surface . . . the scientific outlook knows no insoluble riddles. Analysis of the traditional problems

[1] " Die Wissenschaftliche Weltauffassung. Der Wiener Kreis ", p. 15.

of Philosophy reveals them to be, in part, problems in appearance only (pseudo-problems) and, for the rest, able to be transformed into questions subject to the verdict of the empirical sciences. The clarification of such problems and statements constitutes the object of philosophic activity ".

2. *Relations to Wittgenstein.* Active co-operation is sufficiently rare in Philosophy to deserve attention. More so when the social activity is based on the inspiration of the most solipsist, therefore, by conventional implication, the most anti-social of all philosophers. But if the Viennese circle is very deeply indebted to the opinions of Wittgenstein, it would be a mistake to emphasize that connection so far as to underestimate the considerable influence of Mach and Russell on the circle, or the many elements of novelty owing nothing to either of these predecessors.

The derivation from Mach explains many of the paradoxes involved in the descent of a belligerent group from a philosophy as quietist in temper as

Wittgenstein's[1]. From Mach comes, by direct descent, that belief in the value of group activity which has marked the work of the Viennese circle; from him, also, and by indirect influence through Russell, those elements of Pragmatism already present, by implication, in Wittgenstein, which go far to mitigate the asperities of a marriage between Empiricism and Solipsism. If legitimate, the results of this union are of the highest importance; for it has fallen to many to exorcise Metaphysics from Philosophy, but it has been left for the Logical Positivists, as they are sometimes called, to behave as if they had succeeded. It is not for the translator to estimate their success, or to rob the reader of the pleasures either of invective or applause. But professional philosophers, who have heard with unfailing equanimity their treatises described as compendia of 'nonsense', may be interested to find here detailed and ingenious arguments for refutation; and scientists, who have always found time for a

[1] By Wittgenstein's opinions I mean always to refer to the views expressed in *Tractatus Logico-Philosophicus*. In the absence of any subsequent writings it is not yet possible to say to what extent these views have since been revised.

malicious grin at the expense of Philosophy, are bound to welcome this latest attempt to bring order into a disgraceful muddle of mutually intolerant opinions.

The analytic method adopted by the Viennese circle culminates in the judgment that there are no distinctive philosophical problems. Speculative philosophy must be transformed into a new methodology, the analysis of linguistic forms. But to say so much is to lay too much emphasis on the iconoclastic aspects of these opinions. Their chief concern is to consolidate the achievements of scientific discovery by analyzing the limitations and essential structure of the language in which all knowledge must be expressed. Involved in this programme is the demarcation of boundaries between various departments of linguistic expression, an aim which may receive as much applause from theologians and artists as from scientists, and abuse only from those who find pleasure in the choice of their own labels. Divisions of any sort imply principles of justification ; in the underlying principles on which the Viennese circle base their separation between Sciences and the domains of ' nonsense ' the critic may expect to find their most constructive contributions; here also should be the origin of their divergence from Wittgenstein. For the latter, also, establishes criteria of sense, purges Philosophy of Metaphysics and separates both from the Sciences; but with different consequences.

3. *The notion of sense.* The theory of *Tractatus Logico-Philosophicus* centres round the notion of ' sense ', whose specification is linked with and reveals that conception of the essential structure of language on which is based the doctrine of the untenability of traditional philosophy. Since the sense of statements is defined in terms involving reference to ' atomic ' statements, or ' atomic ' facts, the latter notion is the hub of Wittgenstein's account. The same notion receives a distinct modification in the theories of the Viennese circle.

In the *Tractatus* the world (i.e. the subject-matter of philosophical analysis) is conceived to consist ultimately of simple irreducible ' objects ', occurring in complex arrangements or ' configurations ', and thereby constituting facts or states of affairs. To each ultimate complex arrangements corresponds the simplest kind of true statement, viz., an 'atomic' statement, in such a way as to reflect the form of the fact by a one-one correspondence between objects and words. In virtue of the correlation between constituents of the atomic fact and elements of the true atomic statement, the latter are combined in the same structure as the former; and all statements of more complicated reference to the world than atomic statements must, in so far as they *are* statements and not mere collocations of words, be reducible to logical conjunctions and disjunctions of atomic statements. This doctrine can also be expressed in an alternative and more striking manner: in order to have sense, a statement must be verifiable in (my) experience i.e. the words of which it is composed must be definable in terms of words which refer to (my) *immediate* experience. Unless, indeed, the process of definition ends in this way, in statements composed of words with immediate reference, it will never be possible to know what is meant by the group of words, which will then be just nonsense. Hence the importance of atomic statements; they are the elements on which all other statements are based; all assertions which are not downright nonsense are either truth functions (logical combinations) of atomic statements, or else hypotheses, i.e. rules for constructing atomic statements. Strictly speaking, only atomic statements and truth functions of atomic statements have sense.

In this necessarily inadequate summary of Wittgenstein's teaching three important aspects can be isolated: (I) recognition of the importance of logical structure, (II) the exclusion of logical structure from being itself the subject-matter of statements[1], and (III) verifiability in (my) experience as a necessary criterion of the sense of statements. Easily deducible from these principles is the nonsense of most traditional questions in philosophy, e.g. concerning the existence of other minds, the reality of the external world, etc. Without entering into the detailed criticism which these views deserve, it is possible to point out their dogmatic

[1] I have spoken of statements rather than of facts in accordance with recent practice of the Viennese circle. The whole of the above account can be easily transformed into an 'objective' account in terms of facts, and postulates e.g. no belief in subjective idealism.

character and to hint at their consequences.

It is clear, in the first place, that Wittgenstein's doctrine of nonsense cannot be refuted; for any attempted refutation in other than its own terminology must be ' nonsense ', hence *ipso facto* empty of all assertion; on the other hand, there are no internal inconsistencies to be found in the theoretical formulation of what is essentially a privative doctrine. The most serious consequences of the prohibition of certain combinations of words can be to restrain its defenders, in consistency, from making any assertions at all. That this may in fact be the logical outcome of accepting the system of *Tractatus Logico-Philosophicus* is shown by any attempt to specify in detail the nature of atomic statements (out of which all genuine statements must be constructed). For it has been seen that atomic statements must have immediate reference; hence they must have instantaneous verification. (I can be acquainted with the ultimate ' objects ' only at the moment of acquaintance and reference to them at any other moment, e.g. by description, must be indirect.) Such statements would therefore necessarily be composed of (logically) proper names for the objects of acquaintance during the instant at which the statement is made. Atomic statements would contain no descriptive terms and must therefore be unintelligible except to the speaker and to him only at the moment of utterance; since, moreover, its elements would be merely distinguishable signs for objects of unknown logical form, the atomic statement would be bare of all except the most trivial structure. It is questionable how far such 'statements' have any right to the name; and 'hypotheses', or statements involving any element of generality, are in still worse case, since they are merely prescriptions for the manufacture of atomic statements and can at most be *supported*. Thus atomic statements can be verified but express nothing; while all other statements express but can never be verified.

Consistent adherence to the principles of the *Tractatus* is thus seen to have radical consequences; the philosopher is either to wag his finger, like Kratylos, in lieu of speech, or escapes silence at the expense of discord between theory and practice. Hence the *Tractatus* cannot be said to have succeeded in exhibiting the structure of language and drawing the line between Science and Metaphysics ; for the adopted criterion of sense is so stringent as to exclude the whole of Science from the region of sense in order to share a crowded limbo with the bulk of everyday knowledge and the speculation of metaphysicians.

The absence of detailed discussion of everyday knowledge embodied in ' hypotheses ' is the weakest point of the treatment of the *Tractatus*. Here is where the advantages of the pragmatic approach of the Viennese circle are chiefly to be found. Their modifications (for an adequate account the reader must refer to their own works[1]) can be summarized as the rejection of the substratum of atomic facts while preserving Wittgenstein's insistence of the importance of logical structure. The theory of atomic statements is replaced in Carnap's essay (translated in this book) by the theory of the ' protocol ', the direct record of experience, whose form now remains indeterminate (cf. especially Sections 3, p. 42

[1] E.g. in *Erkenntnis*. Cf. also " Logical Positivism and Analysis " (Henriette Hertz Lecture) by L. S. Stebbing, London, 1933, and Blumberg and Feigl, " Logical Positivism ", *Journal of Philosophy* XXVIII, 281-296 (1931).

below). Rejection of the metaphysical presuppositions of the *Tractatus* is then pursued to the extreme limit of excluding all reference to the 'content' of statements and the practice of a special 'formal'[1] mode of speech from which all such reference has been eliminated. The absence of the doctrine of atomic facts permits the retention of a wider criterion of verifiability in sense experience while allowing a pragmatical sense to general statements, natural laws and hypotheses. This solution brings its own difficulties; it blurs the very definite outlines of the notion of structure in Wittgenstein and leaves truth in an uncomfortable half-way house between correspondence and coherence. For if the truth of statements is provisional some account is needed that does not make their truth dependent on human convenience or human prejudice. This is the difficulty that earlier Pragmatism had to meet and could never answer satisfactorily. For Logical Positivists also, it is a pressing question, but the absence of a final answer cannot detract from the exceptional interest and importance of their work.

[1] Cf. also R. Carnap, "On the Character of Philosophical Problems", *Philosophy of Science*, I, pp. 5-19, 1934.

# AUTHOR'S INTRODUCTION

### THE VIENNESE CIRCLE DOES NOT PRACTICE PHILOSOPHY

The reader may find it easier to understand the main article if I preface it by some remarks on the general nature of the views held by the Viennese Circle to which I and my friends belong.

In the first place I want to emphasize that *we are not a philosophical school and that we put forward no philosophical theses whatsoever.* To this the following objection will be made: You reject all philosophical schools *hitherto,* because you fancy your opinions are quite new; but every school shares this illusion, and you are no exception. No, there is this essential difference, must be the answer. Any new philosophical school, though it reject all previous opinions, is bound to answer the old (if perhaps better formulated) questions. But we give no answer to philosophical questions, and instead *reject all philosophical questions,* whether of Metaphysics, Ethics or Epistemology. For our concern is with *Logical Analysis.* If that pursuit is still to be called Philosophy let it be so ; but it involves excluding from consideration all the traditional problems of Philosophy. In origin, Philosophy included Mathematics and also, until recently, the sciences of Sociology and Psychology. At the present time, these studies have been separated from Philosophy in order to constitute independent branches of Science. Both General Logic and the Logic of Science, i.e. the Logical Analysis of scientific terms and statements, must be separated from Philosophy, in the same fashion, in order to be pursued according to an exact, non-philosophic, and scientific method. Logic is the last scientific ingredient of Philosophy ; its extraction leaves behind only a confusion of non-scientific, pseudo problems.

*Metaphysicians*—whether they are supporters of Monism, Dualism or Pluralism, of Spiritualism, Materialism or some other ' -ism ' propound questions concerning the essence of the Universe, of the Real, of Nature, of History, etc. We supply no new answers but reject the questions themselves as questions in appearance only.

*Epistemology* claims to be a *Theory of Knowledge*, to answer questions as to the validity of knowledge, the basis on which knowledge rests. Here again are to be found many answers from various ' -isms ' ; naive and critical Realism, subjective, objective and transcendental Idealism, Solipsism, Positivism, etc., have as many different answers. We supply no new answer but reject the questions themselves since they seem to have the same character as those of Metaphysics. (The case is altered if the questions are formulated not as philosophical enquiries but as a *psychological* enquiry concerning the origin of knowledge ; in the latter form the question is proper to Science and can be investigated by the empirical methods of Psychology ; but such an answer has nothing to do with the philosophical theses of the -isms mentioned.) If ' Epistemology ' is understood to denote unmetaphysical, purely logical analysis of knowledge, our work certainly falls under that classification.

*Ethics* raises the question of the basis of validity of moral standards (principles of value) and of the specification of valid norms. Answers are given by Idealists, Utilitarians, Intuitionists, etc. Here again we reject the questions themselves in view of their metaphysical character. (The case is otherwise in psychological or sociological investigations of the actions and moral judgments of mankind; such a method is certainly both unobjectionable and scientific, but its results belong to the empirical sciences of Psychology and Sociology, not to Philosophy. It is better to avoid the term 'Ethics' for such investigation in order to avoid confusion with normative or regulative Ethics.)

As against the preceding subjects, our own field of investigation is that of *Logic*. Here are to be found problems of *pure Logic*, i.e. questions relating to the construction of a combined logical and mathematical system with the help of symbolic Logic. Further, the problems of *applied Logic*, or the *Logic of Science*, i.e. the *logical analysis* of terms, statements, theories, proper to the various department of science. Logical Analysis of Physics, for example, introduces the problems of Causality, of Induction, of Probability, the problem of Determinism (the latter as a question concerning the logical structure of the system of physical laws, in divorce from all metaphysical questions and from the ethical question of freedom of will). Logical Analysis of Biology, again, involves the problems of Vitalism, to take one example (but here again in a form free from Metaphysics, viz. as a question of the logical relations between biological and physical terms or laws). In Psychology, Logical Analysis involves, among others, the so called problem of the 'relation between Body and Mind' (here also a non-metaphysical question, concerned not with the essential nature of two realms of being but with the logical relations between the terms or laws of Psychology and Physics respectively). In all empirical sciences, finally, Logical Analysis involves the problem of verification (not as a question concerning the essence of Truth, or the metaphysical basis of the validity of true statements, but as a question concerning the logical inferential relations between statements in general and so called protocol or observation statements).

In this fashion we use Logical Analysis to investigate statements of the various kinds proper to the various departments of Science. The statements of traditional

Philosophy can also be subjected to the same treatment. The result is to reveal the absence of that logical relation (of implication) to empirical statements and, in particular, to protocol statements, whose presence is a necessary condition for the verifiability of the statements in question and is therefore usually, and with justice, required in the findings of all scientific procedure. All statements belonging to Metaphysics, regulative Ethics, and (metaphysical) Epistemology have this defect, are in fact unverifiable and, therefore, unscientific. In the Viennese Circle, we are accustomed to describe such statements as nonsense (after *Wittgenstein*). This terminology is to be understood as implying a logical, not say a psychological, distinction; its use is intended to assert only that the statements in question do not possess a certain logical characteristic common to all proper scientific statements; we do not however intend to assert the impossibility of associating any conceptions or images with these logically invalid statements. Conceptions can be associated with any arbitrarily compounded series of words; and metaphysical statements are richly evocative of associations and feelings both in authors and readers. It is precisely that circumstance which so hinders recognition of their non-scientific character.

In traditional Philosophy, the various views which are put forward are often mixtures of metaphysical and logical components. Hence the findings of the Logical Analysis of Science in our circle often exhibit some similarity to definite philosophic positions, especially when these are negative. Thus, e.g., our position is related to that of *Positivism* which, like ourselves, rejects Metaphysics and requires that every scientific statement should be based on and reducible to statements of empirical observations. On this account many (and we ourselves at times) have given our position the name of Positivism (or New Positivism or Logical Positivism). The term may be employed, provided it is understood that we agree with Positivism only in its logical components, but make no assertions as to whether the Given is real and the Physical World appearance, or *vice versa* ; for Logical Analysis shows that such assertions belong to the class of unverifiable pseudo-statements. Our views are related, in similar fashion, to those of *Empiricism*, since we follow that theory so far as to reject *a priori* judgments ; Logical Analysis shows that every statement is either empirically verifiable (i.e. on the basis of protocol statements), analytic, or self contradictory. On this account, we have at times been classified, both by ourselves and by others, as Empiricists.

The following article is an example of the application of Logical Analysis to investigating the logical relations between the statements of Physics and those of Science in general. If its arguments are correct, all statements in Science can be translated into physical language. This thesis (termed ' Physicalism ' by Neurath) is allied to that of *Materialism,* which respectable philosophers (at least in Germany, whether in other countries also I do not know) usually regard with abhorrence. Here again it is necessary to understand that the agreement extends only as far as the logical components of Materialism ; the metaphysical components, concerned with the question of whether the essence of the world is material or spiritual, are completely excluded from our consideration. In the final section of the article it is shown that methodical Materialism and methodical Positivism are not incompatible ; in the terminology which I have been using here, this is as much as to say : the logical components of Positivism and Materialism are mutually compatible. This same example shows how great is the need for caution in classifying the opinions of the Viennese Circle under any of the old -isms. Between our view and any such traditional view there cannot be identity—but at most agreement with the logical components. For *we pursue Logical Analysis, but no Philosophy.*

Prague, January 1934.                    R. C.

### Advice to the Reader.

Some of the words of most frequent occurrence in the following paper are unfortunately without exact English equivalents; the translations adopted are likely to be misleading without some explanation.

*'Determination'* (for 'Bestimmung') : a description, *or* any indeterminate symbol whose exact value (usually numerical) is obtained as the result of definite operations, *or* the result of such operations.

*Singular* statement (for 'Einzelsatz') : statements describing particular states of affairs in contrast to general statements.

*Physical language* is used technically and does not denote the terminology customary in Physics (cf. p. 95).

*Nonsense* (or *pseudo*-espression) is intended to carry none of its usual abusive connotation. Technical use = whatever cannot be verified in experience.

# PHYSICS AS A UNIVERSAL SCIENCE

By RUDOLF CARNAP

1. THE HETEROGENEITY OF SCIENCE.
2. LANGUAGES.
3. PROTOCOL LANGUAGE.
4. THE PHYSICAL LANGUAGE AS AN INTERSUBJECTIVE LANGUAGE.
5. THE PHYSICAL LANGUAGE AS A UNIVERSAL LANGUAGE.
6. PROTOCOL LANGUAGE AS A PART OF PHYSICAL LANGUAGE.
7. UNIFIED SCIENCE IN PHYSICAL LANGUAGE.

1. THE HETEROGENEITY OF SCIENCE.

Science in its traditional form constitutes no unity, and can be separated into Philosophy and the technical sciences. The latter can be classified again as formal sciences (Logic and Mathematics) and empirical sciences. It is usual to subdivide the last class further and to understand that it includes, in addition to the 'natural' sciences, Psychology and the *Geisteswissenschaften* (social sciences) generally.

The basis of these various divisions is not merely convenience; rather is the opinion generally accepted that the various sciences named are fundamentally distinct in respect of subject matter, sources of knowledge and technique. Opposed to this opinion is the thesis defended in this paper that science is a unity, that all empirical statements can be expressed in a single language, all states of affairs are of one kind and are known by the same method.

Very little will be said here concerning the nature of Philosophy and the formal sciences. The author's views on this point have already been sufficiently explained by others on several occasions. Detailed attention will however be given to the question of the unity of the empirical sciences.

It is to modern developments in logic and particularly in the logical analysis of language that we owe our present insight into the nature of Logic, Philosophy and Mathematics. Analysis of language has ultimately shown that Philosophy cannot be a distinct system of statements, equal or superior in rank to the empirical sciences. For the activity of Philosophy consists rather in clarifying the notions and statements of science. In this way does cleavage of the field of knowledge into philosophy and empirical science disappear; all statements are statements of the one science. Scientific research may be concerned with the empirical *content* of theorems, by experiment, observation, by the classification and organization of empirical material; or again it may be concerned with establishing the *form* of scientific statements, either without regard for content (formal logic) or else with a view to establishing logical connections between certain specific concepts (Konstitutionstheorie and theory of knowledge considered as applied logic).

Statements in Logic and Mathematics are tautologies, analytic propositions, certified on account of their form alone. They have no content, that is to say, assert nothing as to the occurrence or non-occurrence of some state of affairs. If to the statement: "*The (thing) A is black*" we add "*or A is blue*", the supplemented statement still conveys some information though less than at first. If, however, we replace the supplementary phrase previously chosen by ' or

A is not black' the compound statement no longer conveys any information at all. It is a tautology, i.e. is verified by *all* circumstances. From such a statement no knowledge of the properties of the thing A can be derived. Theorems in *Logic* and *Mathematics* have, nevertheless, in spite of tautologous character and lack of content, considerable importance for science by virtue of their use in transforming statements having content. For the present thesis it is important to emphasize that Logic and Mathematics are sciences having no proper subject matter analogous to the material of the empirical sciences. Postulation of 'formal' or 'ideal' objects to be set against the 'real' objects of empirical sciences is unnecessary in the theory here briefly sketched.

Statements having content, i.e. statements, as is usually said, expressing some state of affairs, belong to the field of *empirical sciences*. Our *chief question* is whether these statements, or to speak more conventionally, whether the states of affairs expressed by such statements are divided into several mutually irreducible kinds. The traditional answer is in the affirmative; and it has been usual to make the chief distinctions between the subject matters of natural science, History, the social sciences, etc. (Geisteswissenschaften), and Psychology.

On the basis of observations and experiments, the *natural sciences* describe the spatio-temporal events in the system which we call 'Nature'. From the individual accounts thus obtained arise general formulae, so-called 'laws of nature' (the process of 'induction'). These in turn make it possible to obtain new specific statements, e.g. predictions (the process of 'deduction').

*History, and the social sciences* also use the method of observing material events. The usual view maintains, however, that observation in such fields is merely a subordinate method, the proper method being 'understanding', empathy ('Einfühlung') projection of oneself into historical monuments and events in order to grasp their 'essence'. The further question arises, so it is maintained, in all sciences dealing with culture in the widest sense as well as in specifically normative disciplines such as ethics, of comprehending 'values', of establishing 'norms'. The usual view therefore is that the subject matter of such branches of knowledge, the *Geisteswissenschaften* as Germans say, whether they are 'significant forms' or systems of values, are of a nature fundamentally different from the subject matter of natural science and cannot be understood by the methods of natural science.

As to the nature of *Psychology* widely divergent views are prevalent. Experiments are made, measurements often taken of factors capable of quantitative determination. Many psychologists therefore include their science among the natural sciences, but while doing so accentuate the difference between their respective subject matters. Psychology, they say, deals with the 'psychical', with the phenomena of consciousness, perhaps also of unconsciousness, while other natural sciences treat of the 'physical'. Other psychologists, again, lay the emphasis on the relation between their science and the moral sciences. In Psychology also, they say, knowledge is gained by 'understanding' and empathy. The difference consists in the fact that Psychology does not deal with works of art and institutions, as Ethics and Sociology do, but with the regularities to be found in the phenomena of consciousness. These various conceptions yet agree in the answer they furnish to the questions which we wish to discuss. Psychology is a science with its own fundamentally distinct and isolated subject matter.

We shall not need to discuss in further detail at this point divergent views of the relations between the various sciences. It is sufficient to remember that it is usual to speak of fundamentally distinct kinds of objects ; it matters little for our purpose whether the distinction is made in the manner described above (e.g. ' ideal ' and 'real' objects ; physical, psychical objects ; 'values') or in some other. All such accepted views are contrary to the *thesis of the unity of Science*.

2. LANGUAGES.

In formulating the thesis of the unity of Science as the assertion that objects are of a single kind, that states of affairs are of a single kind, we are using the ordinary fashion of speech in terms of ' objects ' and ' states of affairs '. The correct formulation replaces ' objects ' by ' words ' and ' states of affairs ' by ' statements ', for a philosophical, i.e. a logical, investigation must be an analysis of language. Since the terminology of the analysis of language is unfamiliar we propose to use the more usual mode of speech (which we will call '*material*') side by side with the correct manner of speaking (which we will call the '*formal*'). The first speaks of ' objects ', ' states of affairs ', of the ' sense ', ' content ' or ' meaning ' of words, while the second refers only to linguistic forms.[1]

In order to characterize a definite *language* it is necessary to give its *vocabualry* and *syntax*, i.e. the words which occur in it and the rules in accordance with which (1) sentences can be formed from those words and (2) such sentences can be transformed into other sentences, either of the same or of another language (the so-called

[1] A strictly formal theory of linguistic forms (' logical syntax '), will be developed later. The ' thesis of syntax ' which has only been sketched in the above will there be explained in detail and justified. It asserts that all propositions of philosophy which are not nonsense are syntactical propositions, and therefore deal with linguistic forms. (So-called propositions in metaphysics, on the other hand, can be only the subject-matter of suntactical statements, e.g. of a statement which asserts their syntactical invalidity, i.e. which asserts that they are nonsense.)

(The book here announced is *Logische Syntax der Sprache*, Vienna, 1934.)

rules of inference and rules for translation). But is it not also necessary in order to understand the 'sense' of the sentences, to indicate the 'meaning' of the words? No; the demand thereby made in the material mode is satisfied by specifying the formal rules which constitute its syntax. For the 'meaning' of a word is given either by translation or by definition. A translation is a rule for transforming a word from one language to another, (e.g. 'cheval' = 'horse'); a definition is a rule for mutual transformation of words in the same language. This is true both of so-called nominal definitions (e.g. 'Elephant' = animal with such and such distinguishing characteristics) and also, a fact usually forgotten, for so-called ostensive definitions (e.g. 'Elephant' = animal of the same kind as the animal in this or that position in space-time); both definitions are translations of words.

At the expense of some accuracy we may also characterize a language in a manner other than in the formal mode above and, using the more 'intuitive' material mode, say a language is such that its statements describe such and such (here would follow a list of the objects named in the language). The alternative formulation is permissible provided the writer and the reader are clear that the material mode is only a more vivid translation of the previous description in the formal mode. If this is forgotten the danger may arise of being diverted by the material mode of speech into considering pseudo-questions concerning the essence or reality of the objects mentioned in the definition of a language. Nearly all philosophers and even many Positivists have taken the wrong turning and gone astray in this way.

As an example we may take the language of arithmetic. In the formal mode, this particular language might be characterized as follows :—

Arithmetical statements or sentences are compounded of signs of such and such a kind put together in such and such a way; such and such (specified) rules of transformation apply to them.

Alternatively, using now the material mode, we could say :—

Arithmetical thorems state certain properties of numbers and certain relations between numbers.

Though such a formulation is inexact

it can be clearly understood and is permissible if carefully handled. One must not, however, be led by this formulation into considering pseudo-questions about what kind of objects these 'numbers' are, whether they are 'real' or 'ideal', extra-mental or intra-mental, etc. If the formal mode is used, in which 'numbers' are replaced by 'numerical symbols', such pseudo-questions vanish.

In the rest of the paper we shall at all times help the reader by using both modes of expression and write the formal, and, strictly speaking, only correct, expression of our thought in a parallel column on the left of the more usual formulation.

Various 'languages' can be distinguished in science. Let us for example consider the language of economics which can be characterized in somewhat the following fashion: i.e., by the fact

| that its sentences can be constructed from expressions: 'supply and demand', 'wage', 'price', etc. . . . put together in such and such a way. | that its propositions describe economic phenomena such as supply and demand, etc. |

We will call a language a *universal* language if every sentence can be translated into it, if it can describe every state of affairs, and if this is not the case, a 'partial' language. The language of economics is a 'partial' language since e.g. a theorem in physics concerning the vectors of an electro-magnetic field cannot be translated into the language of economics. the state of an electro-magnetic field in some region cannot be described in economic terms.

3. PROTOCOL LANGUAGE.

Science is a system of statements based on direct experience, and controlled by experimental verification. Verification in science is not, however, of single statements but of the entire system or a sub-system of such statements. Verification is based upon ' protocal statements ', a term whose meaning will be made clearer in the course of futher discussion. This term is understood to include statements belonging to the basic protocol or direct record of a scientist's (say a physicist's or psychologist's) experience.

Implied in this notion is a simplification of actual scientific procedure as if all experiences, perceptions, and feelings, thoughts, etc., in everyday life as well as in the laboratory, were first recorded in writing as 'protocol' to provide the raw material for a subsequent organization. A 'primitive' protocol will be understood to exclude all statements obtained indirectly by induction or otherwise and postulates therefore a sharp (theoretical) distinction between the raw material of scientific investigation and its organization. In practice, the laboratory record of a physicist may have approximately the following form: 'Apparatus set up as follows : . . . . ; arrangement of switches : . . . . ; pointer readings of various instruments at various times : . . . . ; sparking discharge takes place at 500 volts'. Such a set of statements is not a primitive protocol in view of the occurrence of statements deduced with the help of other statements from the protocol. which describe states of affairs not directly observed.

A primitive protocol would perhaps run as follows: "Arrangement of experiment: at such and such positions are objects of such and such kinds (e.g. ' copper wire ' ; the statement should be restricted perhaps to ' a thin, long, brown body ' leaving the characteristics denoted by ' copper ' to be deduced from previous protocols in which the same body has occurred) : here now pointer at 5, simultaneously spark and explosion, then smell of ozone there ". Owing to the great clumsiness of primitive protocols it is necessary in practice to include terms of derivative application in the protocol itself. This is true of the physicist's protocol and true in far greater measure of the protocols made by biologists, psychologists and anthropologists. In spite of this fact, questions of the justification of any scientific statement, i.e. of its origin in protocol statements, involve reference back to the primitive protocol.

From now onwards ' protocol statements ' will be used as an abbreviation for ' statements belonging to the primitive protocol ' ; the language to which such statements belong will be called the ' *protocol-language* '. (Sometimes also termed ' language of direct experience ' or ' phenomenal language ' ; the neutral term ' primary language ' is less objectionable.)

In the present state of research it is not possible to characterize this language with greater precision, i.e. to specify its vocabulary, syntactical forms and rules. This is, however, unnecessary for the subsequent arguments of this paper. The analysis which follows is a sketch of some of the views as to the form of protocol statements held at the present day by various schools of thought. Though the author will take no sides in the issues involved the incidental discussion will elucidate still further the meaning of the term ' protocol-language '.

| | |
|---|---|
| The simplest statements in the *protocol-language* are protocol-statements i.e. statements needing no justification and serving as foundation for all the remaining statements of science. | The simplest statements in the *protocol-language* refer to the given, and describe directly given experience or phenomena, i.e. the simplest states of which knowledge can be had. |
| *Question :* What kinds of word occur in protocol statements? | *Question :* What objects are the elements of given, direct experience? |

*First Answer*: Protocol statements are of the same kind as: 'joy now', 'here, now, blue; there, red'.

*Second answer:* Words like 'blue' do not occur in protocol statements but appear first of all in derived propositions (they are words of higher type). Protocol statements on the other hand are of forms similar to the following :—

(a) 'Red circle, now'

or (b) . . . . .

*First answer:* The elements that are directly given are the simplest sensations and feelings.

*Second answer:* Individual sensasions are not given directly but are the result of isolation. Actually given are more complex objects such as :—

(a) Partial *gestalts* of single sensory fields, e.g. a visual gestalt,

or (b) Entire sensory fields, e.g. the visual fields as a unity,

| or (c) . . . . . . . . | or (c) The total experience during an instant as a unity still undivided into separate sense-regions. |
| *Third answer :* Protocol statements have approximately the same kind of form as ' A red cube is on the table '. | *Third answer :* Material things are elements of the given : a three dimensional body is perceived as such directly and not as a series of successive two-dimensional projections. |

These are three examples of contemporary opinions which are, of course, usually expressed in the material mode. The first can be termed ' Atomistic Positivism ' and is approximately Mach's standpoint. Most present-day critics regard it as inadequate, for objections brought against it by subsequent psychologists and especially followers of Gestalt psychology are to a great extent justified. Opinion on the whole tends to choose between the variations included in the second of the answers given above. The third view in our classification is not often held to-day; it is however more plausible than it appears and deserves more detailed investigation, for which this is however not the place.

Statements of the system constituted by science (statements in the language of that system) are not, in the proper sense of the word, derived from protocol statements. Their relation to these is more complicated. In considering scientific statements, e.g in physics, it is necessary to distinguish in the first place between ' singular ' statements (referring to events at a definite place and time, e.g. ' the temperature was so much at such and such a place and time ') and the so-called ' laws of nature ', i.e. general propositions from which singular propositions or combinations of such can be derived (e.g. ' the density of iron is 7·4 (always and everywhere '). In relation to singular statements a ' law ' has the character of an *hypothesis* ; i.e. cannot be directly deduced from any finite set of singular statements but is, in favourable cases, increasingly supported by such statements. A singular statement (expressed in the vocabulary of the scientific system) has again the character of an hypothesis in relation to other singular statements and in general the same character in its relation to protocol statements. From no collection of protocol statements, however many, can it be deduced, but is in the most favourable case continually supported by them. In fact deduction is possible but in the converse direction. For protocol statements can be deduced by applying the rules of inference to sufficiently extensive sets of singular statements (in the language of the scientific system) taken in conjunction with laws of nature. Now the verification of singular statements consists of performing such deductions in order to discover whether the protocol statements so obtained do actually occur in the protocol. Scientific statements are not, in the strict sense, 'verified' by this process. In establishing the scientific system there is therefore an element of convention, i.e. the form of the system is never completely settled by experience and is always partially determined by conventions.

We will now consider the case of a person A undertaking, with the help of his protocol, verifications of scientific statements in the manner described above. The question whether each person has his own protocol language will be discussed later. For the present A's own protocol language will be referred to as 'the' protocol language.

Whenever the rules of transformation state the conditions under which statements in the protocol language can be deduced from a statement p, it is always possible, in principle, for A to verify p. Whether A can actually do this depends on empirical circumstances. If, however, there is no such inferential relationship between a statement p and statements of the protocol language then p is not verifi- If a state of affairs described by p can be reduced to facts about given, i.e. direct, experience of A, A has in theory the possibility of verifying p. A then knows the 'sense' of p, for the 'sense' of p, or what is expressed by p, consists of the method of verification, i.e. in the reduction to the given. If some statement p is not in this inferential relation to statements concerning the given, p cannot be understood by

able for A ; p has no sense, is formally incorrect. In such a case A cannot understand the statement p, for to 'understand' means to know the consequences of p, i.e. to know the statements of the protocol language which can be deduced from p. If an inferential relation of the kind described holds between a statement p and each of the protocol languages of several persons

A, i.e. p is nonsense.

For if A is to understand a statement he must know what states of affairs involving the given (what possible direct experiences) are the case if p is true.

If the state of affairs expressed by a statement p is verifiable in the manner described by several persons

p has sense for each such person. In such a case p will be said to have sense (for those persons) *inter-subjectively*. This term of course is relative to the persons who understand the statement in the manner described. By a language 'inter-subjective' (for certain persons) will be understood a language whose statements are inter-subjective (for those persons). A statement p, which is inter-subjective (for certain persons), is said to be inter-subjectively valid if p is valid for each person, i.e. if it is supported, in sufficient measure, by the protocol statements of each such person.

It will be proved in the following paragraphs that the physical language is inter-subjective and can serve as a *universal* language, i.e. as a language in terms of which all states of affairs could be expressed. Finally, an attempt will be made to show that the various protocol languages also can be regarded as partial languages, in the sense defined above, of the physical language.

### 4. THE PHYSICAL LANGUAGE AS AN INTER-SUBJECTIVE LANGUAGE.

The physical language is characterized by the fact that statements of the simplest form (e.g. the temperature of such and such a place at a specified time is so much),

attach to a specific set of co-ordinates (three space, one

express a quantitatively determined property of a

time co-ordinates) a definite position at a definite value or a definite time. range of values of a coefficient of physical state.

Quantitative determination can also be replaced by *qualitative*, as is usual in science as well as in everyday life, for reasons of brevity and ease of understanding. Qualitative determinations can therefore be included in the physical language provided rules are set up for translating all such statements into quantitative determinations so that e.g. the statement "It is rather cool here" might be translated into the statement "The temperature here is between 5 and 10 degrees centigrade". they can be understood as determinations of physical states of affairs or occurrences so that e.g. "It is rather cool here" and "The temperature here is between 5 and 10 degrees centigrade" are taken as statements of identical sense.

This characterization of the physical language corresponds to the traditional form of physics (for the sake of simplicity we are neglecting the coefficicnts of probability which occur in the physical statements). We wish however to interpret the term ' the physical language ' so widely as to include not the special linguistic forms of the present merely but also such linguistic forms as physics may use in any future stage of development. It may be that physical position will eventually be determined by more or less than four co-ordinates; perhaps it will not be possible to regard the co-ordinates simply as temporal and spatial magnitudes. Such modifications are of no importance for present purposes. The physical language will certainly continue to be so constituted that every protocol statement composed entirely of words which can be (quite crudely) described as sensation-, perception-, or thing-words, can be translated into it. that every fact of perception in everyday life, e.g. everything that can be learnt about light or material bodies (in the naive interpretation) can be expressed in the physical language.

This property of physical language is sufficient for our further discussion. It is unnecessary to specify further the exact form of physical languages which may possibly arise in the course of the future evolution of physics. In order to facilitate direct understanding the spatio-temporal linguistic forms will always be used in the following paragraphs. On the basis of the property of the physical language just mentioned, our thesis now makes the extended assertion that the physical language is a universal language, i.e. that every statement can be translated into it. every state of affairs can be expressed in it.

In addition to the simplest form of statements previously described, namely of *singular statements*, there are now various compound propositional forms to be considered. The most important is the *general implicative* statement expressing a general implication : if at some point P in space-time there is a determination *a* (i.e. *a* is the value of a certain magnitude at a certain place and time denoted by P) then there will be some other point $P^1$, standing in such and such a spatio-temporal relation to P, at which ($P^1$) there will be with such and such probability a determination $a^1 = f(a)$ functionally dependent on *a*. This is the general form of a *natural law* in its widest meaning. P and P¹ often coincide. An example involving qualitative determinations: 'Blood is red'; a: distinguishing characteristic of blood assumed not to include its colour; $P = P^1$; $a^1$: the colour red. An example involving quantitative determinations: the second of Maxwell's equations:

$$\text{'Curl E} = \frac{\mu}{c} - \frac{\delta H}{\delta t},$$

a: the determination of the spatial distribution of the electrical field in the neighbourhood of P which is denoted by 'curl E'; $P^1 = P$; $a^1$: the rate of change of the magnetic field at P, denoted by $\frac{\delta H}{\delta t}$

The possibility of applying science, i.e. of making predictions concerning subsequent occurrences, depends upon the formulation of laws of nature.

The concepts of physics are quantitative concepts, numerical determinations. This is a fact of decisive importance in permitting prediction on the basis of exact natural laws. Another peculiarity of physical concepts, which is of importance for the present discussion, consists in their abstractness and the absence of qualities from their enunciation. This is to be interpreted as follows: The rules of translation from the physical language into protocol language are of such a kind that no word in the physical language is ever correlated in the protocol language with words referring only to a single sense field (e.g. never correlated with determinations of colour only or sound only). It follows that a physical determination permits the inference of protocol statements in every sensory field. Physical determinations are 'inter-sensory' in a sense which will be immediately explained. Moreover, they are also ' inter-subjective ', in agreement with the experiences of the various subjects; this will be discussed later.

The determination: 'A note of such and such pitch, timbre and intensity ', in the protocol language or in the language of qualities (which we need not at present distinguish) corresponds to the following determination in the physical language : ' Material oscillations of such (specified) basis frequency with superimposed additional frequencies of such (specified) amplitudes'. But a physical statement containing these determinations is correlated not only with statements containing the corresponding determination in the auditory field but also under certain conditions with statements containing determinations from other sensory fields. the presence of such oscillations can be determined not only by auditory sensations (the sound of such a note) but also, with the help of suitable instruments, in the form of visual and tactile sensations.

There are no coefficients of physical state exclusively correlated with quantitative determinations in a single specific sensory field. This is a fact of fundamental importance. For any qualitative determination in some sensory field, we can determine, with the help of qualitative determinations from other sensory fields, the class of the correlated physical determinations. As shown by the illustration used above, qualitative determinations in the auditory field can be translated into physical statements of a particularly simple form. The process is more complicated when determinations of colour are involved, e.g. ' green of such and such a kind ' (denoted by a number, from Ostwald's colour atlas say). Correlated in such a case is not a single physical state but a set of physical states. Every state of this class consists of a definite combination of frequencies of electromagnetic oscillations (e.g. for a definite ' green ' this class includes a combination of a wave-frequency of high intensity from the green part of the spectrum and, a ' red ' frequency of feebler intensity as well as a combination of blue and yellow frequencies of medium intensities, etc.).

It is an important fact that the composition of the set of physical determinations correlated to a qualitative determination can be established experimentally by using the fact that the physical determinations are correlated to qualitative determinations in other sensory fields. Thus, e.g. the composition of the set of combined combinations of frequencies referred to above can be established only in virtue of the fact that the frequencies in question can be recognised by signs other than their respective colours, e.g by the position of the corresponding line in the spectroscopic image. The colours of the spectrum are redundant in the implied experiment since a photograph will furnish all the information required. Hence a person completely blind to colours could still establish frequencies occurring at a definite position in space-time. So far, we have remained inside the region of visual sense, but it is possible to extend this reasoning to other senses. It would be possible for example to build into the spectroscope an electrical apparatus for exploring the spectrum, so constructed that a radiation of sufficient intensity set into motion a pointer which could be felt or a microphone which could be heard. By such means a person completely blind would still be capable of determining the frequency of an electro-magnetic oscillation.

From these arguments follows the theoretical possibility of establishing results of the following three kinds :—

1. *Personal determinations* : A can discover :

which physical determination (or class of physical deter- | under which physical conditions he experiences a definite

minations) corresponds to a definite qualitative determination in his protocol language (e.g. 'green of such and such a kind ').
quality (e.g. a definite *green*).

That determinations of this kind are theoretically always possible is due to the fortunate circumstance (an empirical fact, not at all necessary in the logical sense) that

the protocol     the content of experience.

has certain ordinal properties. This emerges in the fact of the successful construction of the physical language in such a fashion that qualitative determinations in protocol language are single-valued functions of the numerical distribution of coefficients of physical states.

On applying this to our example it follows that the scales of the tactile, visual and photo-spectroscopes can be calibrated in such a fashion that these instruments give the same reading for every given case. In short the same physical determinations correspond to the qualitative determinations of every sensory field; we shall use the abbreviation: *physical determinations are valid inter-sensorily.*

2. *Determinations by other persons.* An experimenter E (e.g. a psychologist) can discover by using another person S (subject of an experiment):

| which physical determination (or set of physical determinations) corresponds to a definite qualitative determination in S's protocol language (e.g. ' green of such and such a colour '). | what are the physical conditions in which S experiences a definite quality (e.g. a definite green). |

The procedure used is the following:— E varies the physical conditions (e.g. the combinations of various frequencies of oscillations) and discovers the conditions to which S reacts with the protocol statement containing the qualitative determination in question. The possibility of such a discovery is independent of

whether the corresponding qualitative determinations (names of shades of colours, etc.) occur in E's protocol language,
whether E can also sense the corresponding qualities,
or of the possibility that E is colour blind or completely blind. For in this case, as in the case of his own experiences, E, as previously stated, receives the same result whether he uses the tactile, auditory or photo-spectroscope. The discovery of the set of these physical determinations corresponding to a definite qualitative determination will be called the '*physicalizing*' or physical transformation of this qualitative determination.

The result of our discussion can now be formulated as follows: A person can physicalize the qualitative determinations both of himself and of another person.

3. *Determinations on other persons made by several experimenters.* If the experiments on a single subject, S, as described in the preceding subsection, are performed not by a single experimenter $E_1$ but by several experimenters $E_1$, $E_2$, ... the various results obtained are in mutual

agreement. This is due to the following fact.

The determined value of a physical magnitude in any concrete case is independent not only of the particular sensory field used but also of the choice of the experimenter. In this we have again a fortunate but contingent fact, viz. the existence of certain structural correspondences between the protocols series of experiences of the various experimenters. A difference of opinion between two observers concerning the length of a rod, the temperature of a body, or the frequency of an oscillation, is never regarded in physics as a subjective and therefore unresolvable disagreement ; on the contrary, attempts will always be made to produce agreement on the basis of a common experiment. Physicists believe that agreement can in principle be reached to any degree of exactitude attainable by single investigators ; and that when such agreement is not found in practice, technical difficulties (imperfection of instruments, lack of time, etc.) are the cause. In all cases hitherto where the matter has been investigated with sufficient thoroughness this opinion has been confirmed. *Physical determinations are valid inter-subjectively.*

Under headings (1) and (3) above we have spoken of a ' fortunate accident ' ; the state of affairs mentioned under (2) is however a necessary consequence of the others. It may be noticed however that these facts, though of empirical nature, are of far wider range than single empirical facts or even specific natural laws. We are concerned here with a perfectly general structural property of experience which is the basis of the possibility of intersensory Physics (fact (1)), and intersubjective Physics (fact (2)), respectively.

The question now arises whether another language exists which is intersubjective and can therefore be considered as a language for Science. The reader's thoughts may turn perhaps to the language of qualities, used say as a protocol language. In virtue of the previously mentioned fact of the possibility of giving a physical interpretation to the language of qualities, the latter must be a sub-language of the physical language. According to customary philosophical opinion, however, there can (or even must) be another non-physical interpretation. It will be shown later that there are objections to such a non-physical interpretation, and that, in any case, the language of qualities, when so interpreted, is not inter-subjective.

It will also be demonstrated that all other languages used in science (e.g. Biology, Psychology or the social sciences) can be reduced to the physical language. *Apart from the physical language* (and its sub-languages) *no intersubjective language is known*. The impossibility of an intersubjective language not included in the physical language has certainly not yet been proved; there are however not the slightest indications to suggest that such a language exists. Further, not a single determination, of any kind, is known which, established intersubjectively, is incapable of translation into the physical language.

It is a just demand that Science should have not merely subjective interpretation but sense and validity for all subjects who participate in it. Science is the system of *intersubjectively valid statements*. If our contention that the physical language is alone in being intersubjective is correct, it follows that *the physical language is the language of Science.*

## 5. THE PHYSICAL LANGUAGE AS A UNIVERSAL LANGUAGE.

In order to be a language for the whole of Science, the physical language needs to be not only intersubjective but also universal. It follows, therefore, to consider whether this is the case, i.e. whether the physical language has the property that

| every statement (whether true or false) can be translated into it. | every possible state of affairs (every conceivable state, whether actually occurring or not) can be expressed by it. |

We will begin by considering the subject matter of the *inorganic sciences,* of Chemistry, Geology, Astronomy, etc. In these regions doubt can hardly arise as to the applicability of physical language. The terminology employed is often different from that of Physics but it is clear that every determination arising can be reduced to physical determinations. For the definition of such determinations is always in terms either of physical determinations or else of qualitative determinations (e.g results of observations); even in the latter case no objections will be raised to the physical interpretation of qualitative observations in these sciences.

The first serious doubts will arise in connection with *Biology*. For the issue of vitalism is still violently controversial at the present time. If we extract the kernel of sense and reject the metaphysical pseudo-questions which are usually confused in the controversy, the essence of the matter in dispute can be stated as the question whether the natural laws which suffice to explain all inorganic phenomena can also be a sufficient explanation in the region of the organic. A negative answer to this question, such as supplied by vitalists, necessitates the formulation of specific and irreducible biological laws. The Viennese circle is of the opinion that biological research in its present form is not adequate to answer the question. We therefore expect the decision to be made in the course of the future development of empirical research. (Meanwhile our presumption tends more to an affirmative answer.) It is however important to notice that the question of the universality of the physical language is quite independent of the vitalist-mechanist controversy. For the former is a question of reducing not biological *laws* to physical laws but biological *concepts* (i.e. determinations, words) to physical concepts. And the fact of the latter reducibility can, in contrast to that of the former, be easily demonstrated. This will perhaps appear obvious immediately the confusion of the two issues is eliminated. Biological determinations involve such notions as species, organisms and organs, events in entire organisms or in parts of such organisms, etc. ; (notions such as ' will ', ' image ', ' sensation ', etc. can be referred to Psychology and omitted from consideration here). Such notions are always defined in Science by means of certain perceptible criteria, i.e. qualitative determinations capable of being physicalized ; e.g. ' fertilization ' is defined as the union of spermatazoon and egg ; 'spermatozoon' and ' egg ' are defined as cells of specified origin and specified perceptible properties ; ' union ' as an event consisting of a specified spatial redistribution of parts, etc. It is possible to define with the help of similar physical determinations the meanings of ' metabolism ', ' cell-division ', ' growth ', ' development ', 'regeneration', etc. The same is true in general of all biological determinations, whose definitions always supply empirical and perceptible criteria.

(This is not the case however, for such words as ' entelechy ', but terms of this sort belong to a vitalistic philosophy of nature rather than to Biology and can occur only in ' nonsensical ' statements. It can be shown that these terms represent pseudo-concepts, incapable of formally correct definition.[1])

The preceding arguments show that every statement in Biology can be translated into physical language. This is true, in the first instance, of singular statements concerning isolated events; the corresponding result for biological laws follows immediately. For a natural law is no more than a general formula used for deriving singular statements from other

[1] Cf. Carnap, " Ueberwindung der Metaphysik durch logische Analyse der Sprache ", *Erkenntnis*, Vol. II, p. 219.

singular statements. Hence no natural law in any field can contain determinations absent from the singular statements in the same field. The question set by vitalism of the relation of biological laws (which laws the foregoing shows to be translatable in all circumstances into physical language and therefore to belong to the general type of physical law) to the physical laws valid in the inorganic realm, does not even arise for consideration here.

The application of our principles to *Psychology* usually provokes violent opposition. In this department of Science, our thesis takes the form of the assertion that all psychological statements can be translated into physical language. This applies both to singular statements and to general statements ('psychological laws'). In other words, the definition of any psychological refer to physical events (viz. physical events in the body, especially the central nervous system, of the person in question) whether of definite single events or in general of events of specified

term reduces it to physical terms. type in a specified person or, more generally still, of such events in any person. In other words, every psychologic concept refers to definite physical properties of such physical events.

The problems raised by these statements are to be dealt with in another paper and will therefore not be discussed further in this place.[1]

If the assertion of the possibility of translating psychological statements into physical language is well grounded, the truth of the corresponding assertion concerning the statements of (empirical) *Sociology* easily follows. Sociology is understood here in its widest sense to include all historical, cultural and economic phenomena; but only the truly scientific and logically unobjectionable statements of these sciences belong to this classification. The sciences mentioned often in their present form contain pseudo concepts,

[1] Carnap, "Psychologie in physikalischer Sprache", *Erkenntnis*, iii, 107-142, 1933.

viz. such as have no correct definition, and whose employment is based on no empirical criteria;

such words stand in no inferential relation to the protocol language and are therefore formally incorrect.

such (pseudo-) concepts cannot be reduced to the given, are therefore void of sense.

Examples: 'objective spirit', 'the meaning of History', etc. By (empirical) Sociology is intended the aggregate of the sciences in these regions in a form free from such metaphysical contaminations. It is clear that Sociology in this form deals only with situations, events, behaviour of individuals or groups (human beings or other animals), action and reaction on environmental events, etc.

These statements may contain physical and also psychological terms. If the foregoing thesis, of the possibility of converting psychological determinations into physical

These events may be in part physical (so-called) and in part mental (so-called). If the foregoing thesis, of the possibility of reducing psychological notions and state-

determinations, is valid, then the same must be true of all sociological terms and statements. ments to physical terms is valid, sociological events must be entirely physical.

These principles were first enunciated in fundamental outline by Neurath,[1] who

[1] Neurath, " Soziologie im Physikalismus ", *Erkenntnis*, Vol. II, cf. also his " Physikalismus ", *Scientia*, Nov. 1931, " Empirische Soziologie. Der wissenschaftliche Gehalt der Geschichte und Nationalökonomie ". *Schriften z. wiss, Weltauff.*, Vol. V, Vienna, 1931. Neurath was also the first both in the discussions of the Viennese circle and, later, in the first article mentioned, to demand constantly the rejection of formulations in terms of ' mental experience '. He rejected the comparison between statements and ' reality ', insisting that the correct mode was in terms only of statements and stated the thesis of physicalism in its most radical form. I am indebted to him for many valuable suggestions. By distinguishing between the ' formal ' and the ' material ' modes, rejecting the pseudo-questions which use of the latter provokes, proving the universality of physical language, and in the consistent application of the formal mode to the construction of syntax (only sketched in the present article) I have arrived at results which wholly confirm Neurath's views. Moreover the demonstration (par. 6), in the present article, that the protocol language can be included in physical language, disposes of our previous difference of opinion on this point (the question of the ' phenomenal language ') which is mentioned in Neurath's article. Neurath's suggestions, which have often met with opposition, have thus shown themselves fruitful in all respects.

has also discussed in detail their bearing on the problems and methods of sociology ; and his papers will be found to include many examples of the possibility of formulation in physical terms and of the elimination of pseudo concepts. We shall therefore be able to omit any further discussion on this point.

The various departments of science have now been inspected. The standpoint of traditional philosophy would demand the inclusion of *Metaphysics*. But logical analysis arrives at the result (cf. f.n. p. 70) that so-called metaphysical statements are no more than pseudo statements,

since they stand in no inferential relation (either positive or negative) to protocol statements. They either contain words irreducible to protocol words or are compounded of reducible words in a manner contrary to the laws of syntax.

since they describe no state of affairs, either existent or non-existent. This is due to the fact that they either contain (pseudo-) concepts which cannot be reduced to the given and therefore denote nothing, or are compounded of sensible concepts in nonsensical fashion.

Our investigations of the various departments of Science therefore lead to the conclusion

| that every scientific statement can be translated into physical language. | that every fact contained in the subject matter of science can be described in physical language. |

We must investigate whether statements in protocol language can also be converted into physical language.

## 6. Protocol Language as a part of Physical Language.

To what extent do statements in protocol language conform to our thesis of the universality of the physical language? That thesis demands that

| statements in protocol language, e.g. statements of the basic protocol, can be translated into physical language. | given, direct experiences are physical facts, i.e. spatiotemporal events. |

Objections will certainly be raised to these assertions. It will be said

> "Rain may be a physical event but not my present memory of rain. My perception of water which is falling at this moment and my present joy are not physical events".

This objection is in the spirit of usual views on this question, and would be accepted by most writers on the Theory of Knowledge. If this objection is considered more closely it will be remarked, in the first place, that it is directed only against the material formulation of our thesis (in the right hand column). We have previously seen that the material mode is a mere transformation of the correct formal mode of speech and easily leads to pseudo-problems. We shall therefore, regard this objection critically in view of the fact that it can be formulated only in the terminology of the right hand column i.e. in the material form, but for the moment, however, we will leave such criticisms on one side and adopt the (fictitious) procedure of regarding the matter from the standpoint of our opponent : we shall, in the first place, use the material mode quite freely and, secondly, suppose that the objection and the grounds on which it is based in its material formulation are justified. It will then appear that we are led into insoluble difficulties and contradictions. This fact will disprove the supposition and dispose of the objection.

Let p be a singular statement in the protocol language of a person $S_1$, i.e. a statement about the content of one of $S_1$'s experiences, e.g. ' I (i.e. $S_1$) am thirsty ' or, briefly, ' Thirst now '. Can the same statement of affairs be expressed also in the protocol language of another person $S_2$? The statements of the latter language speak of the content of $S_2$'s experiences. An experience in the sense in which we are now using the word is always the experience of a definite person and cannot at the same time be the experience of another person. Even if $S_1$ and $S_2$ were, by chance, thirsty simultaneously the two protocol statements ' Thirst now ' though composed of the same sounds would have different senses when uttered by $S_1$ and

$S_2$ respectively. For they refer to different situations, one to the thirst of $S_1$, the other to the thirst of $S_2$. No statement in $S_2$'s protocol language can express the thirst of $S_1$. For all such statements express only what is immediately given to $S_2$; and $S_1$'s thirst is a *datum* for $S_1$ only and not for $S_2$. We do say of course that $S_2$ can 'recognise' the thirst of $S_1$ and can therefore also refer to it. What $S_2$ is actually recognising however is, strictly speaking, only the physical state of $S_1$'s body which is connected for $S_2$ with the idea of his own thirst. All that $S_2$ can verify when he asserts '$S_1$ is thirsty' is that $S_1$'s body is in such and such a state, and a statement asserts no more than can be verified. If by 'the thirst of $S_1$' we understand not the physical state of his body but his sensations of thirst, i.e. something non-material, then $S_1$'s thirst is fundamentally beyond the reach of $S_2$'s recognition.

A statement about $S_1$'s thirst would then be fundamentally unverifiable by $S_2$, it would be for him in principle impossible to understand, void of sense.

In general, every statement in any person's protocol language would have sense for that person alone, would be fundamentally outside the understanding of other persons, without sense for them. Hence every person would have his own protocol language. Even when the same words and sentences occur in various protocol languages, their sense would be different, they could not even be compared. *Every protocol language could therefore be applied only solipsistically ; there would be no intersubjective protocol language. This is the consequence obtained by consistent adherence to the usual view and terminology* (rejected by the author).

But even stranger results are obtained by using, on the basis of our supposition, the material terminology which we regard as dangerous. We have just considered the experiences of various persons and were forced to admit that they belong to completely separated and mutually disconnected realms. We will now consider the relations between the content of my own experiences say, as described by statements in my protocol, and the corresponding physical situation as described by singular statements in physical language, e.g. 'Here the temperature is 20 degrees centigrade now'. We have on the one side the content of experience, sensations, perceptions, feelings, etc., and on the other side constellations of electrons, protons, electro-magnetic fields, etc. ; that is, two completely disconnected realms in this case also. Nevertheless an inferential connection between the protocol statements and the singular physical statements must exist for if, from the physical statements, nothing can be deduced as to the truth or falsity of the protocol statements there would be no connection between scientific knowledge and experience. Physical statements would float in a void disconnected, in principle, from all experience. If, however, an inferential connection between physical language and protocol language does exist there must also be a connection between the two kinds of facts. For one statement can be deduced from another if, and only if, the fact described by the first is contained in the fact described by the second. Our fictitious supposition that the protocol language and the physical language speak of completely different facts cannot therefore be reconciled with the fact that the physical descriptions can be verified empirically.

In order to save the empirical basis of the physical descriptions the hypothesis might perhaps be adopted that although protocol language does not refer to physical events the converse is true and physical language refers to the content of experiences and definite complexes abstracted from such content. Difficulties then arise however on considering the relation between the several persons' protocol languages and physical language. $S_1$'s protocol language refers to the content of $S_1$'s experience, $S_2$'s protocol language to the content of $S_2$'s experience. What can the intersubjective physical language refer to? It must refer to the content of the experiences of both $S_1$ and $S_2$. This is however impossible for the realms of experience of two persons do not overlap. There is no solution free from contradictions in this direction.

We see that the use of the material mode leads us to questions whose discussion ends in contradiction and insoluble difficulties. The contradictions however disappear immediately we restrict ourselves to the correct, formal mode of speech. The questions of the kinds of facts and objects referred to by the various languages are revealed as pseudo-questions. These led us, in turn, to further unanswerable pseudo-questions such as the question how the reciprocal convertibility of physical language and protocol language is compatible with the 'fact' that the first refers to physical situations and the second to experienced content. *These pseudo-questions are automatically eliminated by using the formal mode.* If, instead of speaking of the 'content of experience', 'sensations of colour' and the like, we refer to 'protocol statements' or 'protocol statements involving names of colours' no contradiction arises in connection with the inferential relation between protocol language and physical language. Should then, those expressions in the material mode not be used at all? Their use is in itself no mistake, nor are they senseless, but we see that the danger involved is even greater than previously stated. For complete safety it would be better to avoid the use of the material mode entirely, although it is the terminology usual throughout the whole of Philosophy (also in the Viennese circle). If this mode is still to be used particular care must be taken that the statements expressed are such as might also be expressed in the formal mode. That is the criterion which distinguishes statements from pseudo-statements in Philosophy. [Although the danger that pseudo-questions may arise in using the material mode is always present, the contradictions can be avoided by using the material terminology *monistically*, i.e. by speaking exclusively of the content of experience (in the spirit of solipsism) or else exclusively of physical states (in the spirit of materialism). If, however, a dualist attitude is adopted, as is customary in philosphy, if one speaks simultaneously of ' content of experience ' and ' physical states ', (' matter ' and ' spirit ', ' body ' and ' soul ', ' mental ' and ' physical ', ' acts of consciousness ' and ' intentional objects of consciousness ') then contradictions are unavoidable].

When all contradictions and pseudo-questions have been eliminated by using the formal mode, the problem still remains of analyzing the reciprocal inferential relations between physical language and protocol language. We have previously mentioned that if a sufficient number of physical statements are given, a statement in protocol language can be deduced. A more precise consideration now shows that the simplest form of such deduction is found when physical statements describe the state of the body which belongs to the person in question. All other cases of deduction are more complicated and can be reduced to this case. (In describing the state of the body, the state of the central nervous system and especially the brain is the most important, but further details are unnecessary for our argument.) For example, a protocol statement p : " red now (seen by S) " can be deduced from a definite description of the state of S's body.

The reader may still hesitate, feeling that such a deduction is utopian and would need full knowledge of the physiology of the central nervous system for its performance. This is not however the case ; derivation of the required physical statements is already possible and is achieved in everyday life whenever communication occurs. It is true that what we know in such cases of the physical situation of other persons' bodies cannot as yet be formulated as a numerical distribution of physical coefficients of state but it can be formulated in other expressions of the physical language which are just what we require. Let us, e.g. denote by 'seeing red' that state of the human body characterized by the fact that certain specified (physical) reactions appear in answer to certain specified (physical) stimuli. (For example; Stimulus; the sounds 'What do you see now?' reaction: the sound, 'red'. Stimulus, the sounds, 'Point out the colour you have just seen on this card'; reaction: the finger points to some definite part of the card. Here all those reactions must be counted that are usually regarded as necessary and sufficient criteria for anyone to be 'seeing red now'). It is true that we do not know the numerical distribution of the physical coefficients which characterize the human body in this state of 'seeing red' but we do know many physical events which often occur either as cause (e.g. bringing a poppy before the eyes of the person concerned) or as effect of such a state. (Examples of effects: certain speech-movements; applying a brake in certain situations.) Hence we can first recognize that a human body is in that state and then predict what other states of this body may be expected to occur.

If P be a physical statement: 'The body S is now seeing red '; P is, in the first instance, distinguished from a singular physical statement, in describing not a single point of space-time but an extended spatio-temporal region, viz. the body ; it is further distinguished by corresponding, not to a definite numerical distribution of the coefficients of physical states involved in natural laws, but to a large class of such distributions (whose composition is as yet unknown). If a physical statement is singular in the strict sense, no statement of the protocol language can be deduced from it, nor conversely. But if P is the statement described, the protocol statement p : ' Red (is being seen by S) now ' can be deduced from P and also conversely p from P. In other words p can be translated into P, they both have the same content. (The syntactical concept ' of the same content ' is defined as ' reciprocally inferable '.)

Hence, every statement in the protocol language of S can be translated into a physical statement and indeed into one which describes the physical state of S's body. In other words there is a correlation between S's protocol language and a very special sub-language of the physical language. This correlation is such that if any statement from S's protocol language is true the corresponding physical statement holds intersubjectively and conversely. Two languages isomorphic in this fashion differ only by the sounds of their sentences.

On the basis of this isomorphy we can say *the protocol language is a sub-language of the physical language.* The statement previously made (in the material mode at the time), that the protocol languages of various persons are mutually exclusive, is still true in a certain definite sense : they are, respectively, non-overlapping sub-sections of the physical language. The reciprocal interdependence of the various protocol languages which could not be explained in terms of the previous material account is now seen to be a result of the rules of transformation inside the physical language (including the system of natural laws).

If the result thus obtained, of the

identity of content of P and p, be formulated once again in the two modes, i.e.

" P can be inferred from p, and conversely " 　" P and p describe the same state of affairs "

the material formulation will again provoke the old criticisms. Our previous arguments have prepared us to take a critical attitude towards this formulation. But we will now consider in greater detail the materially formulated objections, for this is the critical point in the argument on which our thesis is based.

Let us assume that $S_2$ makes a report, based on physical observations, of the events in $S_1$'s body yesterday. Then (in the material mode), $S_1$ will not accept this report as a complete account of yesterday's section of his life. He will say that although the report describes his movements, gestures, facial expressions, changes in his nervous system and in other organs it leaves out his experiences, perceptions, thoughts, memories, etc. He will add that these experiences must necessarily be lacking in $S_2$'s report since $S_2$ cannot discover them or at least cannot obtain them by physical observation. Now, we will assume that $S_2$ introduces by definitions, terms such as ' seeing red ' (cf. the example above, p. 86), into the physical language. He can then formulate a part of his report with the help of such expressions in such a way that it runs identically with $S_1$'s protocol. In spite of this $S_1$ will not accept this new report. He will object that although it is true that $S_2$ now uses expressions such as ' joy ', ' red ', ' memory ', etc. he *means* something else than $S_1$ does by the same words in his protocol ; the referends of the expressions are different. For $S_2$, he says, they denote physical properties of a human body, for himself, personal experiences.

This is a typical objection whose form is familiar to all those occupied with the logical analysis of the statements and concepts of Science. If we succeed in demonstrating that some scientific term or other reduces by virtue of its definition to some complex of other determinations and therefore denotes the same as the latter, the objection is always made against us that " we *mean* something else ". If we show that two definite propositions can be deduced one from the other and therefore have the same content, or (in the material form) say the same thing, we hear again and again ' but we *mean* something different when we use the first and when we use the second '. We know that this objection rests upon a confusion between what is expressed by a proposition and the images we associate with the proposition (between ' (logische) Gehalt ' and 'Vorstellungsgehalt') (cf. Carnap: *Scheinprobleme in der Philosophie*).

The same can be said of the present objection. $S_1$ connects different associations with the statements P and p respectively for, on account of their linguistic formulation, P is thought of in connection with physical statements whereas p is associated with the protocol. This difference in associations is however no argument against the thesis that the two propositions have the same content (i.e. express the same), for the content of a proposition is constituted by the possibility of inferring other propositions from it. If the same statements can be inferred from two given statements they must both have the same content, independently of the images and conceptions that we are accustomed to associate with them.

We must now throw more light on the

question of the relation between the protocol statement p to the corresponding physical proposition $P_1$ where both are about physical objects. Let us choose p to be " A red sphere is lying on the table here " and, for $P_1$, " A red sphere (i.e having certain physical properties) is lying on the table ". p has not the same content as $P_1$, for it is possible to have an hallucination of a sphere when there is none on the table, or, conversely, the sphere can be on the table unseen. But p has the same content as another physical statement $P_2$, viz. " S's body is now in physical situation Z ". The situation Z is specified by various determinations including e.g. (1) The stimulus " What do you see ? " is followed by the reaction consisting of the movements, etc. belonging to the sounds ' a red sphere on the table ' ; (2) If a red sphere is laid on the table and S is put in a suitable situation Z occurs. $P_1$ can in certain cases be inferred from $P_2$ ; this necessitates using the definition of Z and suitable natural laws. The argument is from an effect to an habitual cause as used both in Physics and in everyday life. Since $P_2$ can be inferred from p (because they have the same content), $P_1$ can be indirectly inferred from p. The usual interpretation of the protocol statement as referring to a certain condition of the person's environment is therefore an indirect interpretation compounded of the direct reference (to the state of the body) and an appeal to causality.

The conclusion of our discussion is that not only the languages of the various departments of Science but also the protocol languages of all persons are parts of the physical language.

*All statements whether of the protocol, or of the scientific system consisting of a system of hypotheses related to the protocol, can be translated into the physical language. The physical language is therefore a universal language and, since no other is known, the language of all Science.*

7. UNIFIED SCIENCE IN PHYSICAL LANGUAGE.

Our view that protocols constitute the basis of the entire scientific edifice might be termed *Methodical Positivism*. Similarly, the thesis that the physical language is the universal language might be denoted as *Methodical Materialism*. The adjective 'methodical' is intended to express the fact that we are referring to a thesis which speaks simply of the logical possibility of certain linguistic transformations and derivations and not at all of the 'reality' or 'appearance' (the 'existence' or the 'non-existence') of the 'given', the 'mental' or the 'physical'. Pseudo-statements of this kind occasionally occur in classical formulations of Positivism and Materialism. They will be eliminated directly they are recognized as metaphysical admixtures; this is in the spirit of the founders of these movements who were the enemies of all Metaphysics. Such admixtures can be formulated only in the material mode and by eliminating them we obtain Methodical Positivism and Methodical Materialism in the sense defined. When the two views are so purified they are, as we have seen, in perfect harmony, whereas Positivism and Materialism in their historic dress have often been regarded as incompatibles.[1]

Our approach has often been termed 'Positivist'; it might equally well be

[1] Cf. Carnap, *Der Logische Aufbau der Welt*, p. 245 ff.
Frank, "Das Kausalgesetz und seine Grenzen", *Schr. z. wiss. Weltauff.*, Vol. VI, Vienna, 1932, p. 270 ff.

termed 'Materialist'. No objection can be made to such a title provided that the distinction between the older form of Materialism and methodical Materialism— the same theory in a purified form— is not neglected. Nevertheless, for the sake of clarity we would prefer the name of '*Physicalism*'[1]. For our theory is that the physical language is the universal language and can therefore serve as the basic language of Science.

The physicalist thesis should not be misunderstood to assert that the terminology used by physicists can be applied in every department of Science. It is convenient, of course, for each department to have a special terminology adapted to its distinct subject matter. All our thesis asserts is that immediately these terminologies are arranged in the form of a system of definitions they must ultimately refer back to physical determinations. For the sake of precision we might supplement or replace ' physical language ' by the term '*physicalistic language*' ; denoting by the latter the universal language which contains not only physical terms (in the narrow sense) but also all the

[1] Neurath, *loc. cit.*

various special terminologies (of Biology, Psychology, Sociology, etc.) understood as reduced by definitions to their basis in physical determinations.

If we have a single language for the whole of science the cleavage between different departments disappears. Hence the thesis of Physicalism leads to the thesis of the *unity of Science*. Not the physicalist language alone but any universal language would effect a unification of Science but no such language other than the physicalist is known. The possibility of setting up such a language must not, however, be excluded. Its construction would involve the determination of its vocabulary and of its syntax, including rules for transformations inside the language and for inferring protocol statements. Moreover, in accordance with our previous discussion, every proposition P of this language in order to have any sense must allow protocol statements to be inferred according to stated rules. In that case it would be possible, in view of the inferential connection between physical language and protocol language, to construct a statement $P_1$ of the physical language in such a way that all those statements of the protocol language could be inferred from it which could be inferred from P. The two propositions P and $P_1$ of the two different systematic languages would then be so related that in every case where P was true $P_1$ would also be true, and conversely. Hence P could be translated into $P_1$, and conversely.

In general, every statement in the new language could be translated into statements of the physical language and conversely.

every statement in the new language could be interpreted as having the same sense as a statement of the physical language, i.e. every statement of the new language would refer to physical facts, to spatio-temporal events.

Hence, every systematic language of this kind can be translated into the physical language and can be interpreted as a portion of the physical language in an altered dress.

Because the physical language is thus the basic language of Science *the whole of Science becomes Physics*. That is not to be understood as if it were already certain that the present system of physical laws is sufficient to explain all phenomena. It means

| | |
|---|---|
| every scientific statement can be interpreted, in principle, as a physical statement, i.e. it can be brought into such a form that it correlates a certain numerical value (or interval, or probability distribution of values) of a coefficient of state to a set of values of position coordinates (or into the form of a complex of such statements). | every scientific fact can be interpreted as a physical fact, i.e. as a quantitatively determinable property of a spatio-temporal position (or as a complex of such properties). |
| An explanation, i.e. the deduction of a scientific statement, consists of deducing it from a law of the same form as physi- | Every scientific explanation of fact occurs by means of a law, i.e. by means of a formula which express the fact that |

cal laws, i.e. from a general formula for inferring singular statements of the kind specified. situations or events of specified kind in any spatio-temporal region are accompanied by specified events in associated regions related in specified fashion.

It is specifically for *explaining* statements (or facts) by means of laws that a unitary language is essential. It is theoretically always possible inside the total system of Physics

to find an explanation for every singular statement, i.e. a law by means of which this statement (or a corresponding probability statement) can be inferred from other propositions based on the protocol.

to find an explanation for every single fact, i.e. a law in accordance with which this fact is required (with some degree of probability) by the existence of other, known, facts.

For our discussion, it is of no importance whether these laws take the form of unique determinations as assumed in classical Physics (determinism) or, alternatively, as assumed in present day Physics, determine the probability of certain value distributions of parameters (statistical laws of Quantum Mechanics). In contrast to the universality of Physics cases arise in every partial language which can be expressed in that language but are fundamentally incapable of explanation in that language alone, e.g. in Psychology where no explanatory law can be formulated of a statement of the kind "Mr. A is now seeing a red circle" since the explanation must deduce this statement from statements such as "A red sphere is lying before Mr. A" and "Mr. A has his eyes open", etc. e.g. a psychological event such as a perception can be described but not explained; for such an event is conditioned not only by other mental events but also by physical (physiological) events.

The *prediction* of an unknown is similar to the explanation of a known truth or event, viz. derivation with the help of laws. Hence sub- or partial languages are not sufficient for prediction and a unitary language is necessary. If our thesis that there is a unitary language were false, the practical application of Science to most regions would be crippled. It is the fact that physical language supplies the basis for unified Science which first ensures the thorough applicability of Science.

The thesis of *the unity of Science* has nothing to say against the practical separation of various regions for the purposes of division of labour. It is directed only against the usual view that in spite of the many relations between the various regions they themselves are fundamentally distinct in subject matter and methods of investigation. In our view these differences of the various regions rests only upon the uses of various definitions, i.e. of various linguistic forms, of various abbreviations. While

| the statements and words | the facts and objects |

*of the various branches of Science are fundamentally the same kind.* For *all branches are part of the unified Science, of Physics.*

## Classic Works in the History of Ideas

*Also Available in this series:*

ACTION, EMOTION AND THE WILL
*Anthony Kenny*
ISBN 1 85506 319 0 : 1963 Edition : 256pp : £15.99

ARISTOTELIANISM
*John Leofric Stocks*
ISBN 1 85506 222 4 : 1925 Edition : 174pp : £12.99

AUGUSTE COMTE AND POSITIVISM
*John Stuart Mill*
ISBN 1 85506 219 4 : 1865 Edition : 202pp : £14.99

BERKELEY
*G. Dawes Hicks*
ISBN 1 85506 168 6 : 1932 Edition : 346pp : £16.99

BERKELEY – THE PHILOSOPHY OF IMMATERIALISM
*Ian Tipton*
ISBN 1 85506 352 2 : 1974 Edition : 406pp : £16.99

COMMONPLACE BOOK 1919–1953
*George Edward Moore*
ISBN 1 85506 231 3 : 1962 Edition : 426pp : £16.99

DESCARTES
*Anthony Kenny*
ISBN 1 85506 236 4 : 1968 Edition : 256pp : £9.99

THE DIVINE RIGHT OF KINGS
*John Neville Figgis*
ISBN 1 85506 349 2 : 1914 Edition : 420pp : £16.99

ECONOMICS OF INDUSTRY
*Alfred and Mary Paley Marshall*
*New Introduction by Denis O'Brien*
ISBN 1 85506 320 4 : 1879 Edition : 248pp : £16.99

ENGLISH LITERATURE AND SOCIETY IN THE 18TH CENTURY
*Leslie Stephen*
ISBN 1 85506 217 8 : 1904 Edition : 230pp : £12.99

AN ESSAY ON PHILOSOPHICAL METHOD
*R. G. Collingwood*
ISBN 1 85506 392 1 : 1933 Edition : 240pp : £14.99

ESSAYS ON SOME UNSETTLED QUESTIONS OF POLITICAL ECONOMY
*John Stuart Mill*
ISBN 1 85506 160 0 : 1844 Edition : 172pp : £12.99

ESSAYS ON SUICIDE AND THE IMMORTALITY OF THE SOUL
*David Hume,*
ISBN 1 85506 167 8 : 1783 Edition : 132pp : £10.99

ETHICS
*Peter Kropotkin*
*New Introduction by Andrew Harrison*
ISBN 1 85506 224 0 : 1924 Edition : 366pp : £16.99

FOUR DISSERTATIONS
*David Hume*
*New Introduction by John Immerwahr*
ISBN 1 85506 393 X : 1757 Edition : 258pp : £14.99

FRANCIS HUTCHESON
*William Robert Scott*
ISBN 1 85506 169 4 : 1900 Edition : 318pp : £14.99

GOD AND THE SOUL
*Peter Geach*
ISBN 1 85506 318 2 : 1969 Edition : 140pp : £12.99

HOBBES
*George Croom Robertson*
ISBN 1 85506 216 X : 1886 Edition : 250pp : £14.99

HUMAN NATURE, OR THE FUNDAMENTAL ELEMENTS OF POLICY
*bound with*
DE CORPORE POLITICO
*Thomas Hobbes*
*New Introduction by G.A.J. Rogers*
ISBN 1 85506 351 4 : 1840 Edition : 228pp : £14.99

AN INTRODUCTION TO THE PHILOSOPHY OF HISTORY
W. H. Walsh
ISBN 1 85506 170 8 : 1961 Edition : 176pp : £12.99

JOHN STUART MILL
*Alexander Bain*
ISBN 1 85506 213 5 : 1882 Edition : 216pp : £14.99

KANT ON EDUCATION
*Annette Churton (trans.)*
ISBN 1 85506 165 1 : 1899 Edition : 146pp : £10.99

LOCKE
*Samuel Alexander*
ISBN 1 85506 181 3 : 1908 Edition : 102pp : £12.99

LOCKE AND THE WAY OF IDEAS
*John W. Yolton*
ISBN 1 85506 226 7 : 1956 Edition : 248pp : £15.99

MENTAL ACTS
*Peter Geach*
ISBN 1 85506 166 X : 1971 Edition : 148pp : £10.99

OUTLINES OF A PHILOSOPHY OF ART
*R. G. Collingwood*
ISBN 1 85506 316 6 : 1925 Edition : 110pp : £10.99

OF THE CONDUCT OF THE UNDERSTANDING
*John Locke*
*New Introduction by John Yolton*
ISBN 1 85506 225 9 : 1706 Edition : 160pp : £12.99

ON THE AESTHETIC EDUCATION OF MAN, IN A SERIES OF LETTERS
*Friedrich Schiller (Translated by Reginald Snell)*
ISBN 1 85506 322 0 : 1954 Edition : 150pp : £14.99

OUTLINES OF THE HISTORY OF ETHICS
*Henry Sidgwick*
ISBN 1 85506 220 8 : 1886 Edition : 310pp : £14.99

THE PHILOSOPHY OF HEGEL
*G. R. G. Mure*
ISBN 1 85506 237 2 : 1965 Edition : 224pp : £12.99

THE PHILOSOPHY OF KANT
*John Kemp*
ISBN 1 85506 238 0 : 1968 Edition : 138pp : £9.99

THE PHILOSOPHY OF NIETZSCHE
*Abraham Wolf*
ISBN 1 85506 353 0 : 1915 Edition : 120pp : £9.99

PLATO
*David G. Ritchie*
ISBN 1 85506 215 1 : 1902 Edition : 240pp : £12.99

PLATO'S PROGRESS
*Gilbert Ryle*
ISBN 1 85506 321 2 : 1966 Edition : 320pp : £16.99

THE PRESUPPOSITIONS OF CRITICAL HISTORY *bound with*
APHORISMS
*F. H. Bradley*
ISBN 1 85506 214 3 : 1874/1930 Editions : 132pp : £12.99

RELIGION AND PHILOSOPHY
*R. G. Collingwood*
ISBN 1 85506 317 4 : 1916 Edition : 238pp : £14.99

THEORY OF GAMES AS A TOOL FOR THE MORAL PHILOSOPHER
*bound with*
AN EMPIRICIST'S VIEW OF THE NATURE OF RELIGIOUS BELIEF
*R. B. Braithwaite*
ISBN 1 85506 315 8 : 1955 Editions : 116pp : £12.99

THREE ESSAYS ON RELIGION
*John Stuart Mill*
ISBN 1 85506 218 6 : 1878 Edition : 314pp : £14.99

THE UNITY OF SCIENCE
*Rudolph Carnap*
ISBN 1 85506 391 3 : 1934 Edition : 102pp : £9.99

Shimer College
438 N. Sheridan Rd.
P.O. Box A500
Waukegan, IL 60079